Fossilized! PLANT FOSSILS

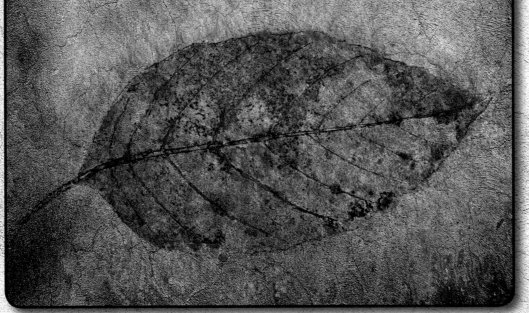

By Kathleen Connors

Gareth Stevens
Publishing

Please visit our website, www.garethstevens.com. For a free color catalog of all our high-quality books, call toll free 1-800-542-2595 or fax 1-877-542-2596.

Library of Congress Cataloging-in-Publication Data

Connors, Kathleen.
Plant fossils / Kathleen Connors.
    p. cm. — (Fossilized!)
Includes index.
ISBN 978-1-4339-6426-8 (pbk.)
ISBN 978-1-4339-6427-5 (6-pack)
ISBN 978-1-4339-6424-4 (library binding)
1. Paleobotany. 2. Plants, Fossil. I. Title.
QE906.C66 2012
561—dc23

                            2011035154

First Edition

Published in 2013 by
Gareth Stevens Publishing
111 East 14th Street, Suite 349
New York, NY 10003

Copyright © 2013 Gareth Stevens Publishing

Designer: Katelyn E. Reynolds
Editor: Kristen Rajczak

Photo credits: Cover, p. 1 Harald Sund/Photographer's Choice/Getty Images; (cover, pp. 1, 3–24 background and graphics), pp. 8–9, 13 Shutterstock.com; pp. 4, 15 Jonathan Blair/National Geographic/Getty Images; p. 5 Chris McGrath/Getty Images; p. 7 Martha Cooper/National Geographic/Getty Images; p. 11 Jeff Foott/Discovery Channel Images/Getty Images; pp. 12, 19 Colin Keates/Dorling Kindersley/Getty Images; p. 16 Paul Zahl/National Geographic/Getty Images; p. 17 Ken Lucas/Visuals Unlimited/Getty Images; p. 20 Gabriel Bouys/AFP/Getty Images; p. 21 Kevork Djansezian/Getty Images.

Printed in the United States of America

CPSIA compliance information: Batch #CW12GS: For further information contact Gareth Stevens, New York, New York at 1-800-542-2595.

CONTENTS

Words in the glossary appear in **bold** type the first time they are used in the text.

ANCIENT PLANTS

Look around a garden on a spring day and you'll see lots of plants beginning to grow. Plants include rosebushes, grasses, and even oak trees! Did you know they **evolved** from ancient plants that date back more than 440 million years?

Scientists have found that the first plants on Earth lived in water. They know this from studying plant fossils. Fossils are the remains or marks of plants and animals that lived thousands or millions of years ago.

This leaf from a sassafras tree has been fossilized.

Many scientists believe land plants evolved from simple green **algae**. However, algae aren't considered plants.

Plant fossils are found all over the world. This is a copy of one that was found in Australia.

PALEOBOTANY

The study of plant fossils is called paleobotany. "Paleo" means ancient. Botany is the study of plants. A paleobotanist often studies one special kind of plant fossil, such as flowers or trees.

One part of paleobotany is palynology, or the study of **spores** and **pollen**. Scientists have studied many of these tiny plant parts because they have been well **preserved** over time. Spore and pollen fossils have helped scientists learn how an area's weather changed over millions of years.

THE FOSSIL RECORD
Plants began to produce pollen about 365 million years ago.

These scientists are studying pictures
of fossil pollen.
▽

7

HOW PLANT FOSSILS FORM

Plant fossils aren't as common as animal fossils. Some of the fossils found most often are pollen, spores, and large, woody plants that don't break down easily. Scientists have found fruit and seed fossils, too. Fossilized flowers, however, are found less often.

Plant fossils form when plants or plant parts are buried in **sediment**. Over thousands or millions of years, the sediment turns to rock and preserves the plants inside. There are many different kinds of plant fossils.

THE FOSSIL RECORD

Plants are often fossilized near bodies of water where there's a lot of moving sediment, such as lakes, rivers, and swamps.

This fern fossil is 300 million years old!

9

COMPRESSIONS AND IMPRESSIONS

Compressions and impressions occur when plant parts become compressed, or flattened, by the great pressure that changes sediment into rock. However, compressions preserve **organic** matter. Impressions don't. They're just prints of the plant parts.

Sometimes, the same plant creates both a compression and an impression. When a rock layer is split apart, the plant matter sticks to one side of it. This side is the compression, or "part." The other side where the plant left a print of itself in the rock is the impression, or "counterpart."

THE FOSSIL RECORD

Both compressions and impressions may clearly show what the outside of a plant looked like. The finer the sediment the plant was buried in, the more clearly these features can be seen.

A compression like this one can tell a lot about the plant's growth and surroundings.
▽

11

CASTS, MOLDS, AND MINERALS

Casts and molds are fossils sometimes created by the same plant, too. After being buried in sediment, plant matter often breaks down, leaving behind a space in the shape of the plant. This is called a mold. When the mold fills with sediment, a fossil called a cast is formed.

Permineralization occurs when **minerals** and water flow into a plant buried in sediment. This mixture fills the empty spaces inside the plant. Sometimes, the minerals take the place of all the plant's parts. This is called **petrification**.

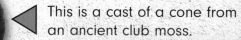

This is a cast of a cone from an ancient club moss.

THE FOSSIL RECORD

In certain types of sediment, plant fossils called compactions are formed. These fossils aren't mineralized but are compacted, or made slightly smaller. Hard fruits are well preserved by compaction.

The most famous petrified plant fossils are petrified forests.

HOW OLD ARE PLANT FOSSILS?

Uncovering plant fossils has helped paleobotanists construct a timeline of plant evolution. They've learned plants didn't live on land until about 440 million years ago. The first forests appeared about 360 million years ago. Scientists think the earliest fossils of plants known as angiosperms date back 100 to 145 million years.

Scientists have many ways of figuring out how old a fossil is. Sometimes they use superposition, or the placement of the rock layer the fossil is in. The higher the rock layer, the younger the fossil is.

THE FOSSIL RECORD

Angiosperms are plants that grow their seeds inside flowers or fruit. Gymnosperms, such as evergreen trees, grow seeds inside cones.

Dating back about 130 million years, this angiosperm fossil is one of the oldest ever found.

15

FLOWERING PLANTS

Today, angiosperms are often called "flowering plants." However, the earliest angiosperm fossils don't show plants with flowers. These plants probably looked more like small bushes or herbs. Some flower fossils date back as far as 120 million years. Plants with big, showy flowers started to grow in large numbers about 95 million years ago.

Fossilized flowers aren't common. Petals break down too easily to be preserved. To learn what these plants were like, paleobotanists use fossilized fruit, seeds, and pollen.

This flower fossil formed when the flower was caught in resin, a sticky matter trees make.

THE FOSSIL RECORD

Some early angiosperms are familiar today. Magnolias, water lilies, and roses started growing about 80 million years ago!

Even without its flower, this lotus leaf fossil can show paleobotanists a lot about the plant it came from.

17

THE ANIMAL CONNECTION

Insect and plant fossils are often found near each other because insects and plants evolved together. Early plants needed help **pollinating**. They grew flowers to draw animals to them. Insects fed from the flowers. They even grew special mouthparts to do so! Insects flew from flower to flower, pollinating as they went.

Plant fossils help scientists learn about the places ancient animals lived. By studying plant fossils found near dinosaur fossils, **paleontologists** can figure out whether the dinosaurs lived in a forest or dry grassland.

THE FOSSIL RECORD

About 60 million years ago, angiosperms started producing fruit and seeds bigger animals wanted to eat. This helped pollination, too.

Fossils like this one show the close link between plants and insects.

▽

19

WHERE TO SEE PLANT FOSSILS

Because many plant fossils are compressions or impressions, it might be hard to recognize them even if there were one in your backyard! However, there are many places to see the fossils of ancient plants.

Museums have collections of plant fossils. The Field Museum in Chicago, Illinois, has about 88,000 plant fossils. Many of these were found in North America. The Florida Museum of Natural History in Gainesville, Florida, has displayed a flowering plant fossil that's 125 million years old.

Some people buy plant fossils and display them in their homes.

PLANT
FOSSIL FACTS

- Fossils from the same plant aren't always found near each other. Paleobotanists often have to put a plant back together in order to study it.

- Plants started to grow on land when a system evolved inside the plants that allowed them to move food and water to all their parts. This is called a plant's vascular system.

- Petrified Forest National Park in Arizona is one of the biggest petrified forests in the world. It's also home to more than 200 different kinds of fossilized plants.

- Flowers are sometimes preserved by wildfires. The organic parts burn away, so the petals don't break down.

GLOSSARY

algae: livings things that grow in water and make their own food, much like plants do

evolve: the process of changes in a living thing that take place over many lifetimes

mineral: matter found in nature that is not living

organic: having to do with living things

paleontologist: a scientist who studies the past using fossils

petrification: the process of organic matter changing to stone

pollen: a fine yellow dust produced by plants

pollinate: to carry pollen to a plant

preserve: to keep safe

sediment: matter, such as stones and sand, that is carried onto land or into the water by wind, water, or land movement

spore: a small body made by a plant that can grow into another plant

FOR MORE INFORMATION

Books

Faulkner, Rebecca. *Fossils.* Chicago, IL: Raintree Publishing, 2007.

Morgan, Sally. *Plant Life Cycles.* Mankato, MN: Smart Apple Media, 2012.

Websites

Finding Fossils
www.sdnhm.org/kids/fossils/
Learn more about all kinds of fossils and where to find them.

Getting into the Fossil Record
www.ucmp.berkeley.edu/education/explorations/tours/fossil/5to8/Intro.html
Read this fun website about how fossils form.

INDEX